AF156899

YOUR KNOWLEDGE HAS VALUE

- We will publish your bachelor's and master's thesis, essays and papers

- Your own eBook and book - sold worldwide in all relevant shops

- Earn money with each sale

Upload your text at www.GRIN.com
and publish for free

Bibliographic information published by the German National Library:

The German National Library lists this publication in the National Bibliography; detailed bibliographic data are available on the Internet at http://dnb.dnb.de .

This book is copyright material and must not be copied, reproduced, transferred, distributed, leased, licensed or publicly performed or used in any way except as specifically permitted in writing by the publishers, as allowed under the terms and conditions under which it was purchased or as strictly permitted by applicable copyright law. Any unauthorized distribution or use of this text may be a direct infringement of the author s and publisher s rights and those responsible may be liable in law accordingly.

Imprint:

Copyright © 2016 GRIN Verlag, Open Publishing GmbH
Print and binding: Books on Demand GmbH, Norderstedt Germany
ISBN: 9783668458420

This book at GRIN:

http://www.grin.com/en/e-book/367145/behaviour-of-salivary-amylase-in-various-reaction-environments-with-reference

Prem Jose Vazhacharickal, Sajeshkumar N.K, Jiby John Mathew, Twinkle Jose

Behaviour of Salivary Amylase in Various Reaction Environments with Reference to Km and Vmax. An Overview

GRIN Publishing

GRIN - Your knowledge has value

Since its foundation in 1998, GRIN has specialized in publishing academic texts by students, college teachers and other academics as e-book and printed book. The website www.grin.com is an ideal platform for presenting term papers, final papers, scientific essays, dissertations and specialist books.

Visit us on the internet:

http://www.grin.com/

http://www.facebook.com/grincom

http://www.twitter.com/grin_com

Behaviour of salivary amylase in various reaction environments with reference to Km and Vmax: an overview

Prem Jose Vazhacharickal, Sajeshkumar N.K, Jiby John Mathew and Twinkle Jose

ACKNOWLEDGEMENTS

Firstly we thank **God Almighty** whose blessing were always with us and helped us to complete this project work successfully.

We wish to thank our beloved Manager **Rev. Fr. Dr. George Njarakunnel,** Respected Principal **Dr. Joseph V.J,** Vice Principal **Fr. Joseph Allencheril**, Bursar **Shaji Augustine** and the Management for providing all the necessary facilities in carrying out the study. We express our sincere thanks to **Mr. Binoy A Mulanthra** (lab in charge, Department of Biotechnology) for the support. This research work will not be possible with the co-operation of many farmers.

We are gratefully indebted to our teachers, parents, siblings and friends who were there always for helping us in this project.

Prem Jose Vazhacharickal*, Sajeshkumar N.K, Jiby John Mathew and Twinkle Jose

Cover page photo courtesy: Wikipedia. Human salivary alpha amylase

Table of contents

Table of figures

Table of tables

List of abbreviations

%	: Percentage
°C	: Degree Celsius
B	: Blank
DNS	: 3, 5-dinitrosalicylic acid
EB	: Enzyme blank
EC	: Enzyme commission
EDTA	: Ethylenediammine tetraacetic acid
g	: Gram
IU	: International unit
Km	: Kinetic constant
L	: Litre
M	: Molar
mg	: Milligram
mL	: Millilitre
mM	: Milimole
N	: Normality
NaCl	: Sodium chloride
NS	: Nelson-Somogyi
SB	: Substrate blank
SD	: Standard deviation
T	: Test
Vmax	: Maximum reaction velocity
μg	: Microgram
μl	: Microlitre
μM	: Micromole

Behaviour of salivary amylase in various reaction environments with reference to Km and Vmax: an overview

Prem Jose Vazhacharickal[1]*, Sajeshkumar N.K[1], Jiby John Mathew[1] and Twinkle Jose[1]

[1]Department of Biotechnology, Mar Augusthinose College, Ramapuram, Kerala, India

Abstract

Amylase is an enzyme which catalyzes the hydrolysis of α (1, 4)-glycosidic linkages in amylose (a linear form of starch), amylopectin (a branched form of starch) and glycogen into simpler carbohydrate molecules such as oligosaccharides or disaccharides. Alpha-amylase is the major form of amylase found in human, most prominently in pancreatic juice and saliva. The salivary amylase is an amylolytic enzyme, which can acts on cooked or boiled starch and converts it in to maltose. So it became interesting to study the behaviour of salivary amylase, when it is secreted as result of different stimuli. And thus began to study the effect of five different stimulatory temperatures, and also the effect of four tastes on the behaviour of salivary amylase. For the study of stimulatory effect of temperature on salivary amylase, five different temperatures are selected (4, 27, 37, 55 and 75°C). And likewise four tastes also selected (sweet, sour, salt and bitter). The DNS method was done in the both tests to obtain the absorbance at 520 nm. The samples were collected from three people, of same age. The saliva was collected at same time, after one and a half hour of their breakfast in order to maintain a controlled condition for this study. In each cases the incubation temperature also kept as variable (4, 27, 37, 55 and 75°C). This study was also aimed to determine the behaviour of salivary amylase with reference to the kinetic parameters like Km and Vmax of salivary alpha amylase by incubating the enzyme (stimulated by different stimulatory conditions of temperature and taste) with varying concentration of substrate. The study revealed the consistency in kinetic parameters like Km and Vmax of salivary alpha amylase secreted in response to various stimuli.

Keywords: Amylase; Vmax; Km; 3, 5-dinitrosalicylic acid method; Environment.

7

1. Introduction

Amylase - one of a group of amylolytic enzyme that catalyzes the hydrolysis of starch into simpler carbohydrate molecules. There are three forms of amylase enzyme. The α-amylase EC 3.2.1.1 occur in saliva, pancreatic juice, malt, and certain bacteria. It is also present in seeds containing starch as a food reserve, and is secreted by many fungi. It catalyzes the hydrolysis of alpha bonds of large, alpha-linked polysaccharides, such as starch and glycogen, yielding glucose and maltose. The β-amylases EC 3.2.1.1 occur in grains, vegetables, malt, and bacteria, is involved in the hydrolysis of starch to maltose. However all amylases are glycoside hydrolases and act on α-1, 4 -glycosidic bonds.

Alpha-amylase is the major form of amylase found in humans and other mammals. Although found in many tissues, amylase is most prominent in pancreatic juice and saliva. Each of which has its own isoform of human α-amylase. They behave differently on isoelectric focusing, and can also be separated in testing by using specific monoclonal antibodies. In humans, all amylase isoforms link to chromosome1p21 (Mandel et al., 2010) Alpha-amylase is one of the major protein components of saliva. Among other proteins, alpha-amylase is synthesized and secreted by acinar cells, after neurotransmitter stimulation (Baum, 1993). Acinar cells are innervated by sympathetic and parasympathetic branches of the autonomic nervous system (Emmelin et al., 1981).

The main function of salivary alpha-amylase is the enzymatic digestion of starch into maltose and dextrin. This form of amylase is also called "ptyalin". Ptyalin acts on linear α (1, 4) glycosidic linkages and, it will break large, insoluble starch molecules into soluble starches (amylodextrin, erythrodextrin, and achrodextrin) producing successively smaller starches and ultimately maltose. Salivary amylase is inactivated in the stomach by gastric acid. In gastric juice adjusted to pH 3.3, ptyalin was totally inactivated in 20 minutes at 37°C. In contrast, 50% of amylase activity remained after 150 minutes of exposure to gastric juice at pH 4.3 (Fried et al., 1987). It is also important for mucosal immunity in the oral cavity, as it inhibits the adherence and growth of bacteria (Bosch et al., 2002).

The salivary amylase gene has undergone duplication during evolution, and DNA hybridization studies indicate many individuals have multiple tandem repeats of the gene. The number of gene copies correlates with the levels of salivary amylase, as measured by protein blot assays using antibodies to human amylase. Gene copy number is associated with apparent evolutionary exposure to high-starch diets. For example, a Japanese individual had 14 copies of the amylase gene (one allele with 10 copies, and a second allele with four copies). The Japanese diet has traditionally contained large amounts of rice starch. In contrast, a Biaka individual carried six copies (three copies on each allele). The Biaka are rainforest hunter-gatherers who have traditionally consumed a low-starch diet. So the variation in starch intake correlates with the number of copies of the salivary amylase gene and amylase saliva in human populations. Increased copy number of the salivary amylase gene may have enhanced survival coincident to a shift to a starchy diet during human evolution (Perry et al., 2007).

The pancreas also makes amylase (alpha amylase) to hydrolyse dietary starch into disaccharides and tri saccharides which are converted by other enzymes to glucose to supply the body with energy. The pancreatic α-amylase randomly cleaves α(1-4) glycosidic linkage of amylose to yield dextrin, maltose, or maltotriose. It adopts a double displacement mechanism with retention of configuration. Genes that are responsible for the production of salivary alpha amylase are AMY1A, AMY1B and AMY1C. The genes responsible for the production of pancreatic amylase are AMY2A and AMY2B (Perry et al., 2007).

Another form of amylase, β-amylase (EC 3.2.1.2) (alternative names: 1, 4-α-D-glucan maltohydrolase; glycogenase; saccharogen amylase) is also synthesized by bacteria, fungi, and plants. Working from the non-reducing end, β-amylase catalyzes the hydrolysis of the second α-1, 4 glycosidic bond, cleaving off two glucose units (maltose) at a time. During the ripening of fruit, β-amylase breaks starch into maltose, resulting in the sweet flavour of ripe fruit.

Both α-amylase and β-amylase are present in seeds; β-amylase is present in an inactive form prior to germination, whereas α-amylase and proteases appear once germination has begun. Cereal grain amylase is key to the production of malt. Many microbes also produce amylase to degrade extracellular starches. Animal tissues do not contain β-amylase, although it may be present in microorganisms contained within the digestive tract. The optimum pH for β-amylase is 4.0-5.0.

Gamma-Amylase (EC 3.2.1.3) (alternative names: Glucan 1, 4-α-glucosidase; amyloglucosidase; Exo-1, 4-α-glucosidase; glucoamylase; 1, 4-α-D-glucanglucohydrolase) will cleave α(1-6) glycosidic linkages, as well as the last α (1-4) glycosidic linkages at the nonreducing end of amylose and amylopectin, yielding glucose. The γ-amylase has most acidic pH optimum because it is most active around pH 3.

1.1 Industrial use of alpha amylase

Alpha and beta amylases are important in brewing beer and liquor made from sugars derived from starch. In fermentation, yeast ingests sugars and excretes alcohol. In beer and some liquor, the sugars present at the beginning of fermentation have been produced by "mashing" grains or other starch sources (potatoes). In traditional beer brewing, malted barley is mixed with hot water to create a "mash," which is held at a given temperature to allow the amylases in the malted grain to convert the barley's starch into sugars. Different temperatures optimize the activity of alpha or beta amylase, resulting in different mixtures of fermentable and unfermentable sugars. In selecting mash temperature and grain-to-water ratio, a brewer can change the alcohol content, mouth feel, aroma, and flavour of the finished beer.

Alpha-Amylase is used in ethanol production to break starches in grains into fermentable sugars. The first step in the production of high-fructose corn syrup is the treatment of cornstarch with α-amylase, producing shorter chains of sugars called oligosaccharides. An α-amylase called "Termamyl", sourced from *Bacillus licheniformis*, is also used in some detergents, especially dishwashing and starch-removing detergents.

1.2 Clinical chemistry and importance

Measurement of serum α-amylase activity is an important diagnostic test for pancreatitis and acute attacks of chronic pancreatitis. Medical laboratories will usually measure either pancreatic amylase or total amylase. The test for amylase is easier to perform than that for lipase, making it the primary test used to detect and monitor pancreatitis. Amylase is measured in Pts with suspected pancreatitis; serum and urine levels peak 4-8 hrs after onset of acute pancreatitis, and normalize within 48-72 hrs; parotitis due to mumps or radiation therapy also increase serum amylase; in cases of increased serum amylase without pancreatitis or parotitis, requires quantification of amylase isoenzymes Ref range Varies by laboratory; 25-90 U/L, serum; 4-30 U/2 hrs, urine; amylase is increase in acute pancreatitis, obstruction of common bile duct, pancreatic duct or ampulla of Vater, pancreatic injury from perforated peptic ulcer and acute salivary gland disease. Amylase is decreased in chronic pancreatitis, pancreatic CA, cirrhosis, hepatitis and eclampsia.

In molecular biology, the presence of amylase can serve as an additional method of selecting for successful integration of a reporter construct in addition to antibiotic resistance. As reporter genes are flanked by homologous regions of the structural gene for amylase, successful integration will disrupt the amylase gene and prevent starch degradation, which is easily detectable through iodine staining.

Salivary α-amylase has been used as a biomarker for stress that does not require a blood draw (Noto et al., 2005). Previous studies revealed that the marked increases in salivary alpha-amylase following psychosocial stress or stress-dependent activation of salivary alpha-amylase (Kirschbaum et al., 1994). The latest research suggests that alpha-amylase is linked to our emotions and our health. In the growing field of amylase research, recent studies have underscored the usefulness of salivary alpha-amylase in this regard. However, some methodological issues have to be resolved in order to integrate salivary alpha-amylase measurements as a standard tool into psycho-physiological research (Rohleder et al., 2006).

1.3 Objectives

The objective of this study was The purpose of this study is to determine whether or not the influence of different stimulatory temperature and tastes correlates with the enzymatic activity of amylase and was studied by the hydrolysis of starch by salivary amylase under various stimulatory conditions (temperature and taste).The activity of salivary amylase is determined by the DNS method (Bernfeld, 1955). The absorbance is read at 520 nm activities in various conditions were calculated from the absorbance using the maltose standard graph.

This study was also aimed to determine the Km and Vmax of salivary alpha amylase by incubating the enzyme (stimulated by different stimulatory conditions of temperature and taste) with varying concentration of substrate. And three incubation temperatures (27°C, 37°C, and 55°C) is setup for each set of reaction in which secretion of salivary alpha amylase is stimulated in response to different stimulatory conditions. By analyzing the obtained Vmax and Km values it is possible to understand the velocity of substrate conversion by the salivary amylase on different reaction conditions and also the affinity of the enzyme with its substrate under various reaction conditions.

And thus analyse the behaviour of salivary alpha amylase secreted in response to various type of stimuli possess any behavioural changes when it undergo in different reaction environments (temperatures).

11

The scope of the present study is to provide, a basic information on the kinetic parameters like Vmax and Km of salivary alpha amylase secreted in response to various stimuli under different reaction environments, for development of enzyme systems in vitro for various applications of research and commercial importance in future.

2. Review of literature

2.1 Salivary alpha amylase

Amylase is a calcium dependent enzyme which hydrolyzes complex carbohydrates at α (1, 4)-glycosidic linkages to form maltose and glucose. Amylase is a major component of human saliva, plays a role in the initial digestion of starch. Foods that contain much starch but little sugar, such as rice and potato, taste slightly sweet as they are chewed because amylase turns some of their starch into sugar in the mouth. It may also be involved in the colonization of bacteria involved in early dental plaque formation. Salivary amylase also possesses a suitable site for binding to enamel surfaces and provides potential sites for the binding of bacterial adhesins.

The pancreas also makes amylase (alpha amylase) to hydrolyse dietary starch into disaccharides and trisaccharides which are converted by other enzymes to glucose to supply the body with energy. More surprisingly, α-amylase is also found in blood, sweat and tears, possibly for anti-bacterial activity Plants and some bacteria also produce amylase. Specific amylase proteins (isoforms of enzyme) are designated by different Greek letters (α, β, γ –amylase).All amylases are glycoside hydrolases and act on α-1, 4-glycosidic bonds.

The α-amylases (EC 3.2.1.1) (alternative names: 1, 4-α-D-glucan glucanohydrolase; glycogenase) are calcium metalloenzymes, completely unable to function in the absence of calcium. By acting at random locations along the starch chain, α-amylase breaks down long-chain carbohydrates, ultimately yielding maltotriose and maltose from amylose, or maltose, glucose and "limit dextrin" (oligo saccharide) from amylopectin. Because it can act anywhere on the substrate, α-amylase tends to be faster-acting than β-amylase. In animals, it is a major digestive enzyme, and its optimum pH is 6.7-7.0.

α-Amylase   α-Amylase

Starch ⟶ Oligosaccharides ⟶ Maltose + Glucose

In vitro, α -Amylase is also able to hydrolyze the α (1,4) linkages in glycogen, but has no activity on the α (1,6) linkages responsible for the more highly branched structure of glycogen. These branched structures also reduce the activity of α-Amylase toward glycogen by limiting the accessibility of the target α (1, 4)-glucan bonds.

The amount of amylase present in saliva varies with the composition of the diet and is present in largest amounts when the diet contains a large amount of the carbohydrates that are hydrolyzed by this enzyme. The salivary amylase levels found in the human lineage are six to eight times higher in humans than in chimpanzees, which are mostly fruit eaters and ingest little starch relative to humans.

The first enzyme that the food encounters in our mouth is called salivary amylase. It is released by salivary glands and is the most abundant enzyme in saliva (Ramasubbu et al., 1996). There are three main pairs of salivary glands - the parotid glands, the submandibular glands (also called the submaxillary glands) and the sublingual glands. In addition, there are between six hundred and a thousand minor salivary glands located in the mouth, throat and lips.

The parotid glands are the largest salivary glands. One parotid gland is located in each cheek, in front of the ear. The parotid glands produce a watery liquid containing protein. The two submandibular glands (or submaxillary glands) are located under the floor of the mouth. These glands produce a liquid that is a mixture of water and mucus. The two sublingual glands are located under the tongue, in front of the submandibular glands, produce a liquid that contains more mucus than the secretions of the other salivary glands. Saliva leaves the glands in tubes called salivary ducts. More saliva is made when the mouth contains spicy, sour or acidic foods. When taste buds are stimulated by these chemicals they trigger the release of saliva.

Saliva is thick, colourless and glistening liquid consisting of about 98% to 99% water. It kills bacteria, helps to prevent tooth decay, begins the digestion of food, helps to speak and helps to swallow food as well. Saliva contains many chemicals in addition to water, including mucus, salts, antibacterial substances (lysozyme, lactoferrin, peroxidase and immunoglobulin A), enzymes, other proteins and buffering agents (Sodium bicarbonate) to keep the pH at the correct level in the mouth. It also contains bacterial cells, since bacteria live in our mouths, and human cells shed by the lining of the mouth, the tongue and the gums. The salivary glands generally make between one and two litres of liquid a day (between two and four pints). During an average lifetime they produce about 10,000 gallons of saliva. Saliva is released continuously from the salivary glands, although the amount varies during the day. The quantity increases when taste, smell or even think of food, as well as when we eat certain kinds of food, but it decreases during the time of sleep.

Saliva contains an enzyme called salivary amylase or ptyalin, which digests starch into a sugar called maltose. (Maltose is later broken up into glucose molecules in the small intestine). When chewing, teeth break down the food into physically smaller pieces that can be acted on by digestive enzymes. The first enzyme that the food encounters in our mouth is called salivary amylase. It is released by our salivary glands and is the most abundant enzyme in our saliva (Ramasubbu et. al, 1996).

2.2 Alpha amylase assay
Several methods are available for determination of α-amylase activity, and different industries tend to rely on different methods. The starch iodine test, a development of the iodine test, is based on colour change, as α-amylase degrades starch and is commonly used in many applications. A similar but industrially produced test is the Phadebas amylase test, which is used as a qualitative and quantitative test within many industries, such as detergents, various flour, grain, and malt foods, and forensic biology. The Nelson-Somogyi (NS) and 3, 5-dinitrosalicylic acid (DNS) assays for reducing sugars are widely used in measurements of carbohydrase activities against different polysaccharides.

Maltose can be used as a standard for estimating reducing sugar in unknown samples. Constructing a standard curve / graph for maltose helps us to estimate concentration of reducing sugars present in an unknown sample and for determining the activity of amylase enzyme in forthcoming experiments. The standard curve for maltose is usually constructed using DNS as the reagent. Maltose reduces the pale yellow coloured alkaline 3, 5-Dinitro salicylic acid (DNS) to the orange- red coloured, 3 amino, 5 nitro salicylic acid.

Saccharolytic activity of alpha -amylase was measured with the DNS method (Bernfeld, 1955). The 3, 5-Dinitrosalicylic acid (DNS or DNSA, IUPAC name 2-hydroxy-3, 5-dinitrobenzoic acid) is an aromatic compound that reacts with reducing sugars and other reducing molecules to form 3-amino-5-nitrosalicylic acid, which absorbs light strongly at 540 nm. The DNS procedure uses 1 % soluble starch as substrate.100 µl of the enzyme were incubated for 30 min at 37°C with 2.5ml of phosphate buffer (0.02 M, pH 7.1) and 2.5ml of soluble starch. A blank without substrate but with α-amylase extract and a control containing no α-amylase extract but with substrate were run simultaneously with the reaction mixture. The reaction was stopped by addition of 0.5 ml of DNS and heated in boiling water for 5 min prior to read absorbance at 540 nm. One unit of α-amylase activity was defined as the amount of enzyme required to produce 1 mg of maltose in 30 min at 37 °C.

A standard curve of absorbance against amount of maltose released was constructed to enable calculation of the amount of maltose released during α-amylase assays. Amylase activity can be defined in international (IUB) units of micromoles of product/min per litre of saliva. It was first introduced as a method to detect reducing substances in urine and has since been widely used, for example, for quantifying carbohydrates levels in blood. It is mainly used in assay of alpha-amylase. However, enzymatic methods are usually preferred due to DNS lack of specificity (Miller, 1959).

The study about the activity of salivary amylase, it role and concentration in saliva are now a day's widely subjected to the various fields of study and research works. The salivary amylase is now used as the biomarker for stress. Its concentration variation in the saliva is considered as disease conditions. A large number of studies are reported about the salivary amylase day by day. And most of them are about the relations between stress and increase of salivary amylase concentration. 'The psychosocial stress-induced increase in salivary alpha-amylase is independent of saliva flow rate is the one kind of study' done by Rohleder et al. (2006) and Kirschbaum et al. (1994). They reported that flow rate is not confounder of stress-induced salivary amylase activation.

Mehrabadi and Bandani (2009) assayed the α-Amylase activity based on Bernfeld method by the DNS procedure. The activity of α-amylase in salivary glands was 0.050 U/insect. The optimum pH and temperature for the enzyme activity was determined to be 6.5-7 and 30-35°C, respectively. The enzyme activity was inhibited by addition of EDTA (Ethylenediamine tetraacetic acid) urea, $CaCl_2$, $MgCl_2$ and SDS but Mg^{2+}, NaCl and KCl enhanced enzyme activity.

Ramasubbu et al. (1996) conclude that aromatic residues in the vicinity of the active site of human salivary α-amylase play a crucial role in substrate binding, enzyme activity and catalysis. Perry et al. (2007), did a study on 'Diet and the evolution of human amylase gene copy number variation' They found that copy number of the salivary amylase gene (AMY1) is correlated positively with salivary amylase protein level and that individuals from populations with high-starch diets have, on average, more AMY1 copies than those with traditionally low-starch diets.

3. Hypothesis

The current research work is based on the following hypothesis

1) Temperature and environmental stimulus affect the activity of salivary amylase.
2) Vmax and Km value of salivary amylase varies with stimulus.

4. Materials and Methods

4.1 Study area

Kerala state covers an area of 38,863 km^2 with a population density of 859 per km^2 and spread across 14 districts. The climate is characterized by tropical wet and dry with average annual rainfall amounts to 2,817 ± 406 mm and mean annual temperature is 26.8°C (averages from 1871-2005; Krishnakumar et al., 2009). Maximum rainfall occurs from June to September mainly due to South West Monsoon and temperatures are highest in May and November.

4.2 Sample collection

The samples were collected from three healthy individuals with same age. The samples were collected after one and a half hour, during the intake of food. Samples were collected by giving respective stimulatory condition (temperature and taste). Prior to collecting sample one minute is allowed for the salivation in response to the stimulatory condition provided. Then immediately removed the stimulatory substance and saliva was collected by mixing with 2 ml of distilled water by the time of salivation. Then this 2ml sample was mixed with 23 ml distilled water, thus made a 1/25 dilution sample. For the temperature as the stimulatory condition, the 2ml distilled water with respective degree of temperature was given in to mouth, and the sample was collected immediately after one minute from the stimuli given. And this 2 ml sample was mixed with 23 ml distilled water, and thus made a 1/25 dilution sample which was used for the enzyme assays.

4.3 Reagents

4.3.1 Phosphate Buffer (0.2 M; pH- 7)
Solution A: Prepare 31.2 gm of Dihydrogen sodium phosphate (156.1 MW) in1000 ml distilled water.
Solution B: Prepare 28.392 gm of Disodium hydrogen phosphate (102 MW) in 1000 ml distilled water.
Working buffer: Prepared by mixing 170 ml solution A with 152.2 ml solution B, (pH.7).

4.3.2 Maltose standard solution
Prepare 100 mg maltose solution in a standard flask and made up to100ml with distilled water (1 mg/ml).

4.3.3 DNS Reagent
Prepare 1 gm of 3, 5-dinitrosalicylic acid (DNS) in 30 gm of sodium potassium tartarate and dissolved in 20 ml of 2N NaOH. This solution was made up to 100ml with distilled water and stored in an amber coloured bottle.

4.3.4 1% starch solution
Prepare 1 gm starch in 100 ml distilled water, and dissolved the solution by heating.

4.3.5 1% NaCl solution
Prepare 1 gm NaCl in 100 ml distilled water, and dissolved the solution by mixing.

4.3.6 2N NaOH
Prepare 8 gm NaCl in 100 ml distilled water, and dissolved the solution by mixing.

4.3.7 Enzyme solution

Prepared by in taking 2 ml of distilled water in to mouth and wait for 1minute then mix this 2 ml saliva with 23 ml distilled water and stored in a cooled condition.

4.4 Assay procedure

4.4.1 Preparation of maltose standard graph

1. Pipette out standard maltose solution in the range of 0.6, 1.2, 1.8, 2.4 and 3 ml, into 5 separate test tubes.
2. A test tube containing a blank solution is also prepared.
3. Using distilled water, bring the volume up to 3 ml in each test tube, including the test tube containing the blank solution.
4. Add 3 ml of DNS reagent to each tube and cover the test tubes with aluminium foil.
5. Heat the contents in the test tubes in a boiling water bath for 10 minutes.
6. Cool the test tubes to room temperature, after taking them out of the water bath.
7. Then add 1 ml distilled water to each test tube and mix well.
8. Take 3 ml from each test tube into different cuvettes and place each cuvette in a colorimeter and record the intensity of dark orange red colour at 520 nm as the 'absorbance' or OD.
9. Plot a graph with the amount of maltose on X axis Vs OD at 520 nm on Y axis.

4.4.2 Assay of α –amylase

1. Set up 4 clean and dry test tubes, labelled as blank 'B', enzyme blank 'EB', substrate blank 'SB' and test 'T'.
2. Pipette out 2.5 ml of 0.2M phosphate buffer (pH 7.0) into all test tubes.
3. Add 2.5 ml of 1% starch solution to test tube labelled as 'EB' and 'T' and mix well.
4. Add 1 ml of 1% NaCl solution to all the four tubes, mix well, and then incubate all the tubes at room temperature for 5 min.
5. Pipette out 2.5 ml of distilled water to test tube labelled as 'B' and 'SB'.
6. Add 0.1ml of diluted saliva to the tube labelled as 'SB' and 'T' and mix well.
7. Then add 0.1 ml of distilled water to test tubes labelled as 'B' and 'EB', mix well and incubate at 37°C for 15 min in a water bath.

8. Add 0.5 ml of 2N NaOH to all the tubes, after taking them out of the water bath and mix well.

9. Then add 0.5 ml of DNS to all the tubes, mix well and incubated them in boiling water bath for 5min. Cool the test tubes to room temperature, after taking them out of the water bath.

10. Take 3 ml from each test tube into different cuvettes and place each cuvette in a colorimeter and record the intensity of dark orange red colour at 520 nm as the 'absorbance' or OD and the activity of salivary amylase is calculated.

Concentration of maltose (in moles) liberated is calculated using the formulae;

$$\text{Moles of maltose} = \frac{\text{weight in gms/ml} \times \text{volume taken} \times 10^{-3}}{\text{Molecular weight of maltose}}$$

4.4.3 Determination of Km value

1. Set up twelve – clean and dry test tubes and labelled as $T_1, C_1, T_2, C_2, T_3, C_3, T_4, C_4,$
   T_5, C_5 and $T_6, C_6.$

2. A test tube containing a blank solution is also prepared.

3. 0.5 ml of 0.5%, 1%, 1.5% 2%, 2.5% and 3% substrate solution was added in to $T_1, C_1, T_2, C_2, T_3, C_3, T_4, C_4, T_5, C_5$ and T_6, C_6 respectively.

4. Add 1 ml of 1% NaCl solution to all tubes, mix well, and then add 4.ml of phosphate buffer (pH 7.0)to the tubes labelled as blank 'B'.

5. Pipette out 1.5 ml of phosphate buffer (pH 7.0) in to $T_1, C_1, T_2, C_2, T_3, C_3, T_4,$ C_4, T_5, C_5 and $T_6, C_6.$ Mix well and incubate at 37°C for 5 min in a water bath.

6. Add 0.1 ml of diluted saliva to the tube labelled as T_1, T_2, T_3, T_4, T_5 and $T_6.$ Mix well and add 0.1 ml of distilled water into C_1, C_2, C_3, C_4, C_5 and $C_6.$

7. Pipette out 2 ml 0f distilled water to all the test tubes, except blank. Mix well and incubate at 37°C for 15 min in a water bath.

8. Add 1 ml of 2N NaOH to all the tubes, after taking them out of the water bath and mix well.

9. Then add 1 ml of DNS to all the tubes, mix well and incubated them in boiling water bath for 5min. cooled the test tubes to room temperature, after taking them out of the water bath.
10. The optical density was measured at 520 nm against reagent blank.
11. The amount of maltose liberated was calculated using the standard graph of maltose.

The velocity of reaction in each test tube was calculated by using the formula;

$$\text{Velocity} = \frac{\text{Amount of maltose liberated (µg)}}{\text{Incubation time}}$$

4.5 Statistical analysis

The survey results were analyzed and descriptive statistics were done using SPSS 12.0 (SPSS Inc., an IBM Company, Chicago, USA) and graphs were generated using Sigma Plot 7 (Systat Software Inc., Chicago, USA).

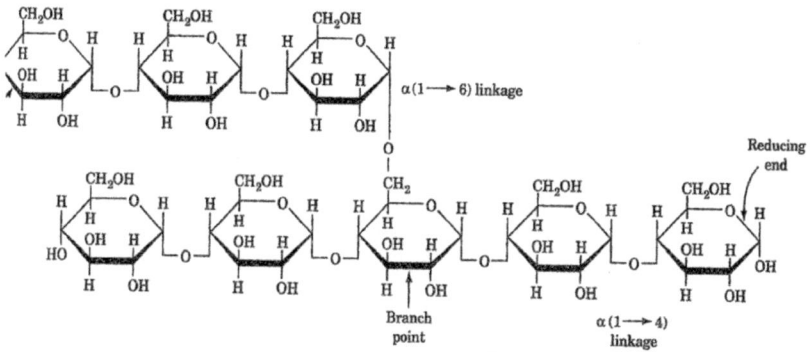

Figure 1. Sugar backbone linkage.
Photo courtesy: Wikipedia. https://en.wikipedia.org/wiki/Glycosidic_bond

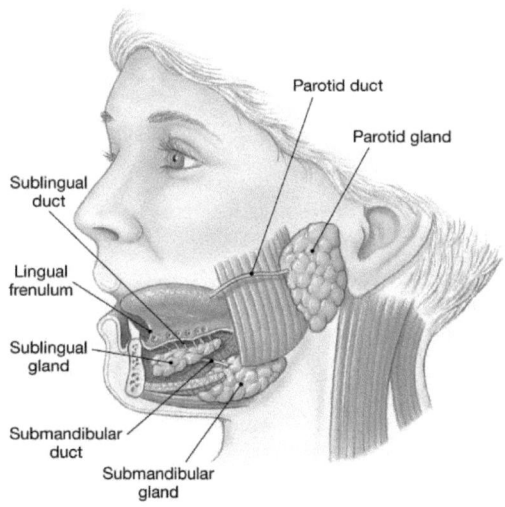

Figure 2. Description of the salivary glands in human mouth. Photo courtesy: Simplyknowledge.com.

http://www.simplyknowledge.com/uploads/gknowledge/salivary-glands/salivary-glands.jpg

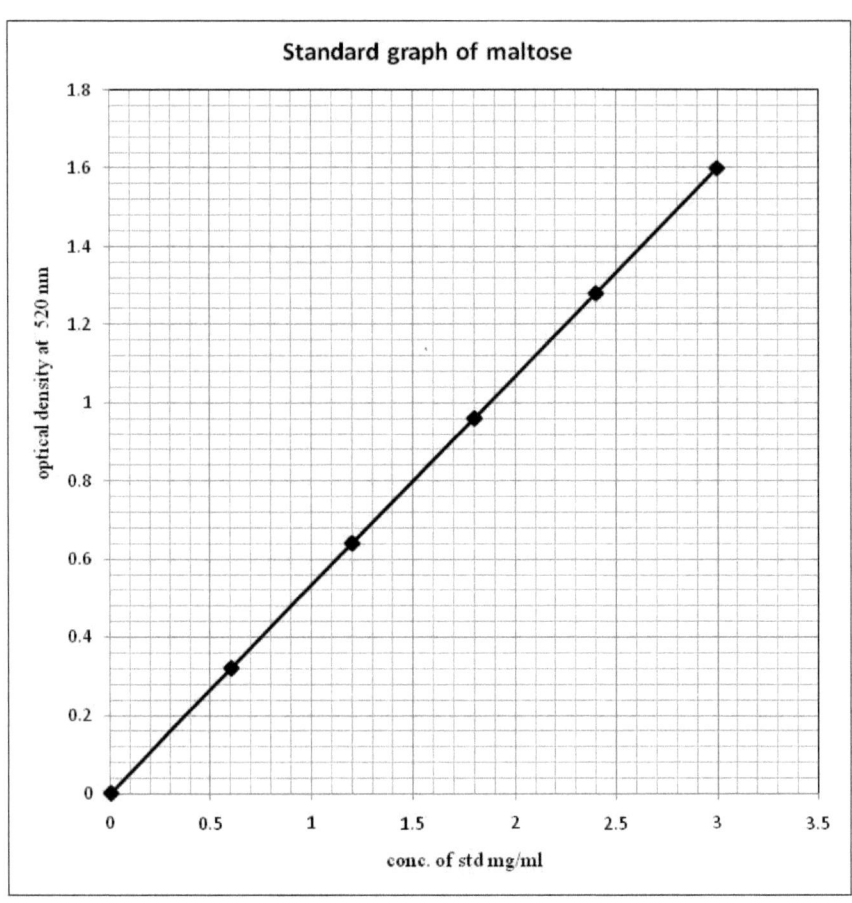

Figure 3. Description of the standard graph of maltose. Authors own image

Table 1. Protocol for the construction of standard graph of maltose

	B	S_1	S_2	S_3	S_4	S_5
Volume of standard solution (ml)	_	0.6	1.2	1.8	2.4	3.0
Concentration of standard solution (mg/ml)	_	0.6	1.2	1.8	2.4	3.0
Volume of distilled water (ml)	3.0	2.4	1.8	1.2	0.6	_
Volume of DNS (ml)	3	3	3	3	3	3
Mix well and kept in boiling water bath for 10min.						
Volume of distilled water (ml)	1	1	1	1	1	1
Optical density at 520 nm	0	0.32	0.64	0.96	1.28	1.60

Table 2. Protocol for the determination of Km and Vmax of salivary amylase.

Serial no	Blank	Enzyme blank	Substrate blank	Test
0.2 M phosphate buffer (ml)	2.5	2.5	2.5	2.5
1% Starch (ml)	____	2.5	____	2.5
1% NaCl (ml)	1	1	1	1
Incubate at room temperature for 5 minutes				
Distilled water (ml)	2.5	____	2.5	____
Diluted saliva (ml)	____	____	1	1
Distilled water (ml)	1	1	____	____
Incubated at Respective Temperature for 15 minutes				
2N NaOH (ml)	0.5	0.5	0.5	0.5
DNS Reagent (ml)	0.5	0.5	0.5	0.5
Incubate at boiling water bath for 5 minutes				
OD at 520 nm	___	___	___	___

Table 3. Activity of salivary α- Amylase stimulated by different temperature.

	Sample	Stimulation Temperature (°C)					Incubation temperature (°C)
		4	27	37	55	75	
Activity of salivary amylase in undiluted saliva (value ×10⁻⁴ µ moles/sec)	Sample-1	0.0515	0.0509	0.0532	0.0555	0.0491	4
		0.0555	0.0480	0.0538	0.0555	0.0509	27
		0.0416	0.0543	0.0520	0.0480	0.0538	37
		0.0515	0.0515	0.0532	0.0543	0.0567	55
		0.0497	0.0509	0.0515	0.0555	0.0486	75
	Sample-2	0.0555	0.0462	0.0543	0.0543	0.0497	4
		0.0555	0.0538	0.0515	0.0555	0.0503	27
		0.0410	0.0520	0.0526	0.0515	0.0532	37
		0.0532	0.0491	0.0503	0.0549	0.0561	55
		0.0538	0.0486	0.0532	0.0520	0.0486	75
	Sample-3	0.0532	0.0480	0.0555	0.0555	0.0515	4
		0.0538	0.0538	0.0543	0.0578	0.0491	27
		0.0410	0.0509	0.0532	0.0509	0.0526	37
		0.0503	0.0543	0.0543	0.0543	0.0572	55
		0.0549	0.0491	0.0503	0.0538	0.0509	75

Table 4. Activity of salivary α- Amylase stimulated by different taste.

	Sample	Stimulation taste				Incubation Temperature (°C)
		Salt	Sweet	Sour	Bitter	
Activity of salivary amylase in undiluted saliva (value ×10⁻⁴ µ moles/sec)	Sample-1	0.0497	0.0538	0.0520	0.0515	27
		0.0538	0.0567	0.0555	0.0486	37
		0.0526	0.0555	0.0509	0.0538	55
	Sample-2	0.0567	0.0555	0.0538	0.0555	27
		0.0543	0.0555	0.0561	0.0486	37
		0.0520	0.0538	0.0497	0.0474	55
	Sample-3	0.0526	0.0567	0.0532	0.0526	27
		0.0543	0.0538	0.0555	0.0486	37
		0.0520	0.0526	0.0520	0.0503	55

Table 5. Sample mean and Standard deviations of the Km for different samples of

			Km value of Samples			Mean value of samples	Standard deviation
			Sample-1	Sample-2	Sample-3		
Stimulaion Factors	Temperature (°C)	27	0.5	0.52	0.5	0.50	0.01
		37	0.52	0.52	0.52	0.52	0
		55	0.50	0.47	0.50	0.49	0.01
	Taste	Salt	0.47	0.50	0.50	0.49	0.01
		Sweet	0.50	0.50	0.52	0.50	0.01
		Sour	0.50	0.55	0.55	0.53	0.02
		Bitter	0.50	0.525	0.47	0.51	0.02

salivary- amylase.

Table 6. Sample mean and Standard deviations of the Vmax for the different samples of salivary- amylase.

			Vmax of Sample			Mean value of samples	Standard deviation
			Sample-1	Sample-2	Sample-3		
Stimulaion Factors	Temperature (°C)	27	176	172	176	174.66	2.30
		37	192	190	192	191.33	1.15
		55	192	192	192	192.00	0
	Taste	Salt	194	192	192	192.66	1.15
		Sweet	192	192	196	193.33	2.30
		Sour	192	190	192	191.33	1.15
		Bitter	192	192	190	191.33	1.15

5. Results and discussion

The influence of different stimulatory temperatures and tastes on kinetic properties of salivary alpha amylase was studied by DNS assays for reducing sugars that reacts with reducing sugars and other reducing molecules (that are produced by hydrolysis of starch by salivary amylase under various stimulatory conditions of temperature and taste) to form 3-amino-5-nitrosalicylic acid, which absorbs light strongly at 520 nm. The analyses of the behaviour of salivary alpha amylase secreted in response to various types of stimuli are carried out by two assays. The first assay involves the determination of the activity of salivary amylase that is stimulated in response to the different stimuli given. The next assay was to determine the Km and Vmax of the enzyme in the same conditions that are provided in the first assay and the behavioural changes are analysed from the obtained result.

A standard graph is plotted using the optical density obtained from the assay for the standard graph of maltose. The samples were collected from three healthy individuals with same age. The samples were collected after one and a half hour after the intake of food. Samples were collected by giving respective stimulatory condition of taste and temperature, prior to sample collection and one minute is allowed for the secretion of saliva in mouth in response to the stimulatory condition provided. Saliva was collected in to 2 ml of distilled water during salivation. Then the 2 ml sample was mixed with 23 ml distilled water, thus made a 1/25 dilution sample. This sample was stored in a cooled condition and used for the assays.

The sample mean and standard deviation in the kinetic parameters like Km and Vmax of salivary amylase, obtained at the different stimulatory conditions. When 27°C was given as the stimulus for salivation, the sample mean of Km is 0.5083 mg% with standard deviation 0.0144.In the next experiment 55°C used as the stimulus for salivation, and the same deviation is obtained, with the sample mean of Km 0.491 mg%', even if the Vmax of 27°C (174.66 ± 2.309 μg) shows a great difference from the Vmax of 55°C (192 ± 0 μg). When comparing deviation of Km in 55°C and 27°C with the Km obtained from 37°C as the stimulus for salivation, the standard deviation of Km (0.525 mg%) with 37°C as stimulus (0) is nil, but the Km at which 37°C was used as stimulus shows a large variation from Km in 55°C and 27°C of stimuli. And also it shows standard deviation of ±1.154 in Vmax (191.33 μg), from the mean value compared to the standard deviation of Vmax obtained from 27°C as stimulus. From the above obtained data of temperature it is very clear that the salivary amylase stimulated by 37°C shows a comparatively high substrate affinity than the other two stimulatory temperature, even if a smaller difference with the Vmax of saliva stimulated by 55°C and 27°C.

Though the Vmax represent the velocity of enzyme action and Km is the substrate affinity at half maximal velocity, the salivary amylase possess comparatively a high affinity to substrate and also with comparatively high velocity of enzyme reaction, when it stimulated by 37°C, than the other two temperature used for this study.

And in the case of tastes are used as stimuli for saliva secretion, taste such as salt, sweet, and bitter, shows the same standard deviation (± 0.01443), from the mean value of Km which are same as to standard deviation of Km value obtained when 27°C and 55°C were used as the stimuli. But the standard deviation of Km obtained when sour (0.5333 ±0.0288mg %) was used as the stimuli, shows a small and considerable difference with standard deviation obtained from other tastes. And the standard deviations of Vmax of obtained from salt, sour and bitter are also same (192.66 ± 1.1547 μg).But the standard deviation of Vmax obtained when sweet (193.33 ± 2.309 μg) was used as the stimuli, shows a small difference from standard deviation nof Vmax obtained from other tastes. By analyzing the obtained data of activity of salivary amylase stimulated by various tastes, it was observed that the salivary amylase shows a relatively high Vmax (193.33 ± 2.3094 μg), when it is stimulated by sweet taste and lower Km compared to sour taste used as stimulus for saliva secretion.

Salivary amylase possesses comparatively high activity (iVmax) when sour and sweet tastes were used as stimuli for the salivation, with in this the sour taste shows the maximum activity. The substrate affinity is comparatively lower in case of bitter as the stimulus for saliva secretion. The samples that are stimulated in different temperature possess its maximum activity when the same temperature of stimulation is given as the incubation temperature. In the case of taste, maximum activity is shown (in the order) by sweet, sour, salt and bitter. When extreme temperatures such as 75°C and 4°C were given for the stimulation for salivation and also as incubation temperature during the assay, there was no significant difference in the activity of the saliva stimulated in normal temperature.

6. Conclusions

An interesting fact observed in the study is that irrespective of samples, stimulatory conditions and reaction environments; value of the Vmax maintained a constant level (191.33) with a narrow range of deviations ± 2.309) and at the same time the Km value is more consistent; the SD of Km is less when compared with that of Vmax. The stimulus has a small influence in the activity of salivary amylase with reference to their Km and Vmax. Even though the Vmax of salivary amylase at different stimuli are different, the Km value is maintained in all cases with a narrow range of deviation.

From these results it can be concluded that activity and affinity to the substrate of salivary amylase is maintained consistently with narrow range of deviations during various stimulatory conditions. As the expression of salivary amylase is genetically controlled and the secretion is a response to the stimuli, the change in activity of a protein such as salivary amylase is very much related to its structure also. So it can be assumed that the consistent activities during different environmental conditions are due to the expression of its allels to form isoforms in response to various types of stimuli.

More studies on the genetic, molecular and biochemical mechanisms related the expression of salivary amylase in response to various environmental conditions, can supplement to validate the findings of the present study.

Acknowledgements

The authors are grateful for the cooperation of the management of Mar Augusthinose college for necessary support. Technical assistance from Binoy A Mulanthra is also acknowledged.

References

Baum, B. J. (1993). Principles of saliva secretion. *Annals of the New York Academy of Sciences, 694*(1), 17-23.

Bernfeld, P. (1955). [17] Amylases, α and β. *Methods in enzymology, 1*, 149-158.

Bosch, J. A., Ring, C., de Geus, E. J., Veerman, E. C., & Amerongen, A. V. N. (2002). Stress and secretory immunity. *International Review of Neurobiology, 52*(1), 213-253.

Emmelin, N. I. L. S., Young, J. A., & Garrett, J. R. (1981). Nervous Control of Mammalian Salivary Glands. *Philosophical Transactions of the Royal Society of London B: Biological Sciences, 296*(1080), 27-35.

Fried, M, Abramson, S, Meyer, J. H. (1987). Passage of salivary amylase through the stomach in humans. *Digestive Diseases and Sciences 32*(1), 1097-1103.

Ghalanbor, Z., Ghaemi, N., Marashi, S. A., Amanlou, M., Habibi-Rezaei, M., Khajeh, K., & Ranjbar, B. (2008). Binding of Tris to *Bacillus licheniformis*??-Amylase can affect its starch hydrolysis activity. *Protein and Peptide Letters, 15*(2), 212-214.

Kirschbaum, C., Scherer, G., & Strasburger, C. J. (1994). Pituitary and adrenal hormone responses to pharmacological, physical, and psychological stimulation in habitual smokers and nonsmokers. *Journal of Molecular Medicine, 72*(10), 804-810.

Lide, David R. (1998). Handbook of Chemistry and Physics (87 ed.). Boca Raton, FL: CRC Press. pp. 3–318.

Mandel, A. L., des Gachons, C. P., Plank, K. L., Alarcon, S., & Breslin, P. A. (2010). Individual differences in AMY1 gene copy number, salivary α-amylase levels, and the perception of oral starch. *PloS One, 5*(10), e13352.

Mehrabadi, M., & Bandani, A. R. (2009). Study on salivary glands α-amylase in wheat bug *Eurygaster maura* (Hemiptera: Scutelleridae). *American Journal of Applied Sciences, 6*(4), 555-560.

Miller, G. L. (1959). Use of dinitrosalicylic acid reagent for determination of reducing sugar. *Analytical Chemistry, 31*(3), 426-428.

Noto, Y., Sato, T., Kudo, M., Kurata, K., & Hirota, K. (2005). The relationship between salivary biomarkers and state-trait anxiety inventory score under mental arithmetic stress: a pilot study. *Anesthesia & Analgesia, 101*(6), 1873-1876.

Perry, G. H., Dominy, N. J., Claw, K. G., Lee, A. S., Fiegler, H., Redon, R., & Carter, N. P. (2007). Diet and the evolution of human amylase gene copy number variation. *Nature Genetics, 39*(10), 1256-1260.

Ramasubbu, N., Paloth, V., Luo, Y., Brayer, G. D., & Levine, M. J. (1996). Structure of human salivary α-amylase at 1.6 Å resolution: implications for its role in the oral cavity. *Acta Crystallographica Section D: Biological Crystallography, 52*(3), 435-446.

Ramasubbu, N., Paloth, V., Luo, Y., Brayer, G. D., Levine, M. J. (1996). "Structure of human salivary α-Amylase at 1.6 Å resolutions: implications for its role in the oral cavity". *Acta Crystallographica Section D Biological Crystallography, 52*(3), 435–446.

Rohleder, N., Wolf, J. M., Maldonado, E. F., & Kirschbaum, C. (2006). The psychosocial stress-induced increase in salivary alpha-amylase is independent of saliva flow rate. *Psychophysiology, 43*(6), 645-652.

Voet, D., & Voet, J. G. (2005). Biochimie. (2e éd.). Bruxelles: De Boeck. 1583 p.

YOUR KNOWLEDGE HAS VALUE

- We will publish your bachelor's and
  master's thesis, essays and papers

- Your own eBook and book -
  sold worldwide in all relevant shops

- Earn money with each sale

Upload your text at www.GRIN.com
and publish for free